The Messerschmitt Me P.1101

David Myhra

Schiffer Military History
Atglen, PA

Book Design by Ian Robertson.

Copyright © 1999 by David Myhra.
Library of Congress Catalog Number: 99-61085.

Printed in China.
ISBN: 0-7643-0908-0

We are interested in hearing from authors with book ideas on military topics.

Published by Schiffer Publishing Ltd.
4880 Lower Valley Road
Atglen, PA 19310 USA
Phone: (610) 593-1777
FAX: (610) 593-2002
E-mail: Schifferbk@aol.com.
Visit our web site at: www.schifferbooks.com
Please write for a free catalog.
This book may be purchased from the publisher.
Please include $3.95 postage.
Try your bookstore first.

In Europe, Schiffer books are distributed by:
Bushwood Books
6 Marksbury Road
Kew Gardens
Surrey TW9 4JF
England
Phone: 44 (0)181 392-8585
FAX: 44 (0)181 392-9876
E-mail: Bushwd@aol.com.

Try your bookstore first.

Messerschmitt Projekt 1101

The *Me P.1101* was *Willy Messerschmitt's* labor of love...his favorite single turbojet-powered in house research project. He wanted to believe that the *1101* would allow *Messerschmitt AG* to exceed *Mach 1* (the speed of sound) with a fighter aircraft in level flight before any other aircraft manufacturer in the world! But according to its designer W*oldemar Voigt* the *1101* was an idea that simply wouldn't "jell." *Messerschmitt AG* entered the *1101* in the *RLM's* "*Emergency Fighter Competitions*" in late 1944 but lost out to the *Focke-Wulf Ta 183* design of *Hans Multhopp*. With the machine's rejection *Messerschmitt* gave it another purpose...it would be used to test the flight characteristics of varying degrees of wing sweptback while flying...that is with in a range of 20° to 45°s of sweep back in order to deal favorably with the sound barrier or sonic barrier. The mechanics of changing the *1101's*

degree of wing sweep while flying had not yet been fully worked out but the *1101s* wing swept could be pre-fixed manually on the ground in steps of 35°, 40°, and to a maximum of 45 degrees.

Messerschmitt AG started on the preliminary design study for a high performance single turbojet- powered interceptor, having a highly swept-back wing about July, 1944. The basic design following *RLM* specifications included:

- 1 1/2 hours endurance at approximately 32,800 feet [10,000 meters].
- A single *HeS 011A* turbojet engine producing 2,866 pounds [1,300 kilograms] of thrust.
- Fuel supply—about 370 gallons [1,400 liters].
- 106 mph [170 km/h] stalling speed at a landing weight plus _ fuel and full ammunition.

- 2x MK *108 30* mm cannon minimum with 3 or 4xM*K 108 30* mm cannon preferred.

The *P.1101* was discovered by American Army troops as they entered the Tyrol Mountain area in Bavaria around Oberammergau south of Munich. *Messerschmitt AG's* chief designer *Woldemar Voigt* of the Advanced-Design Drawing Office (ADDO) had been working on the *1101's* design off and on since mid 1942. It is reported that when the American Army discovered the *1101* in its workshop that its wings were not mounted to the fuselage and the machine was estimated to be about 80% flight ready. Workers interrogated at the *Messerschmitt* facility post war thought that the *1101's* maiden flight was scheduled for June, 1945. Later its wings were attached by American Army Engineers. It was about this time, too, that the *1101* was moved outdoors on the grass to serve as a static display along with several other *Luftwaffe* aircraft including a *Me 262*. The *262* had no engines and had been a hangar mate to the *1101*. It is not known why the engine-less *262* was in the same workshop as the *P.1101*. But once outdoors in the soft ground the *1101* fell back on its tail section and its port side main wheel gear collapsing as it sank down into the earth. When it was relieved of its display duties 12 months later and taken to *Bell Aircraft*, Buffalo, New York, the *1101* was missing parts, seriously deteriorated from the weather, and its skin full of dents, the result of American soldiers walking, sitting, and otherwise sliding around on the machine and having their pictures taken for the folks back home in the United States.

The proposed *Me P.1101 fighter version in camouflage and* seen from its nose starboard side. Scale model by *Mike Hernandez*. Photographed by *Tom Trankle*.

To see his privately funded aeronautical research treated by the Americans like a piece of playground equipment then shipped off to the United States for detailed analysis, and when the experts had finished, discarded in a metal scrap heap, angered *Willy Messerschmitt* the rest of his life. Nor did he ever forgive his chief designer *Voigt* for his decision to accompany the *1101* to America in 1946 and even worse...deciding to stay. *Messerschmitt* remained a *National Socialist* in belief to the day of his death and as was his life-long habit seemed to blame others for his failure. Consequently, he blamed *Voigt* as the reason for the *1101's* failure to win *RLM* approval, funding and being turned into a weapon of war. Post war *Messerschmitt* bitterly called *Voigt* a traitor nor did he have any love for the United States. But all that was different back in Autumn, 1944 when *Oberst Siegfried Knemeyer* of the *RLM* issued a request for bids on a single *HeS 011A* turbojet powered, high altitude fighter with better performance than the current *Me 262* fighter/fast bomber with its twin *Jumo 004B* turbojet engines. *Messerschmitt's P.1101* was one of eight proposed designs submitted to the *RLM*. Other proposals submitted included proposed designs from *Blohm und Voss*, *Focke-Wulf*, *Heinkel*, and *Junkers*. The second generation *HeS 011A* turbojet engine promised 2,866 pounds [1,300 kilograms] of thrust compared to the *Jumo 004B* with its 1,984 pound [900 kilogram] thrust.

As aircraft research projects go, *Messerschmitt AG's P.1101* had a short of rugged simplicity about it. American aviation experts found the machine rather ugly and since it was originally as a fighter but redesigned as a research aircraft to test sweep back wing degrees in the sonic barrier with no provisions for guns, radar, and combat-related equipment, military experts regarded the *1101* as having only minimal intelligence value.

As early as 1933 *Voigt* had been interested in theoretical work on swept wings for aircraft done by *Dr.Ing. Adolf Busemann*, one of Germany's most respected aerodynamitists. Later *Dr.Ing. Albert Betz* of the *Deutsches Versuchsanastalt für Luftfahrt (DVL)* or German Experimental Institute for Aviation continued research into the relationship between compressibility and wing sweptback in high speed flight. This continuing research was followed by *Voigt* and others in the German aviation industry, indeed, aviation industries world-wide. As the research work into minimizing air compressibility at *DVL* showed increasing promise, *Messerschmitt* came to sponsor a comprehensive wind tunnel program to test the soundness of *Busemann's* aerodynamic theories with special emphasis on aircraft wing design. Based on the findings of *Busemann's* research *Voigt* and *Messerschmitt* about July, 1942 started primary design for a flying machine with a sweptback wing increments of 30° to 40° suggested by B*usemann's* on going research at *DVL*. This flying machine came to be known as *Projekt 1101*. Ideally *Voigt* and *Messerschmitt* wanted to be able to fix the wing sweep at one angle, fly it, re-position it, fly it, and so on. This way the two men could see for themselves how the aircraft performed throughout its entire performance envelop and then reset the wing and do it all over again in order to make comparison. But before *Busemann's* research findings were applied to a full scale *1101* airframe *Voigt* wanted to make a scale model. His *1101* scale model had a wing span of 6.6 feet [2 meters] and it was extensively tested at *DVL's* Berlin-Adlershof wind tunnel with encouraging results. By this time, too, the *Me 262* was being flight tested. Originally it not had been designed with swept back wings but with a straight wing. The wings became swept back in order to maintain the center of gravity as the airframe went to holding the heavier *Jumo 004s* from the first engines of choice...the smaller diameter and lighter *Bramo/BMW 002* turbojet engines. One of the findings of the *262's* flight testing was that a single engine aircraft such as the *1101* would require a great deal more thrust than the *004* was capable of providing and a more powerful 2nd generation turbine such as the *HeS 011A* was not anywhere ready for field testing. So in 1942 work on *Projekt 1101* slowed considerably for the lack of a German turbojet engine with sufficient thrust. Moreover, since the *262* was requiring more work than had

A slightly overhead view of the *P.1101* fighter version from its nose port side. Scale model in camouflage paint by *Mike Hernandez*. Photographed by *Tom Trankle*.

been anticipated (and because of *Hitler's* misguided demand to have the *262* fitted out as a fast bomber then back to a fighter, and so on) *Voigt* and other members of the *Messerschmitt* advanced-design team had to devote their full attention to that aircraft before its purpose as a war machine was finally settled.

During the critical days of the M*e 262* some design work was still being carried out on *Projekt 1101*. This is evidenced by the many different design drawings researchers have located all claiming to be *Projekt 1101*. The *1101's* inactive status was changed about September, 1944 and after two years of relative idleness. This change in attitude came about mainly when the *RLM's* department of technical development under *Knemeyer* issued a requirement for a single-engine fighter to be powered by the 2nd turbojet engine...the more powerful 2,866 pound thrust *HeS 011A*. *Heinkel-Hirth* officials had told the *RLM* that this engine would be ready for serial production in early 1945. It never was but in September, 1944, *Messerschmitt AG* offered the *RLM* their *P.1101* as the right aircraft to be used against American *B-17* bomber formations...and the high altitude *B-29s* which were expected, too. *Projekt 1101's* competition included B*lohm und Voss' P.215*, *Focke- Wulf's P.183*, *Heinkel's P.1078A*, and *Junkers' EF 128*. The *RLM* did not make an award but several companies were encouraged to construct a prototype of their entry...*Messerschmitt AG* among them. To help these companies get on with the production of a prototype, the *RLM* provided a small amount of funds.

As the P*rojekt 1101* prototype was taking shape at Oberammergau under bomber less skies (the Allies did not learn of M*esserschmitt AG's* extensive aviation research facilities located there until post war) *Voigt* struggled with construction the prototype which seemed unwilling to gell into a winning project. The principle difficulties confronting *Voigt* and others at *Messerschmitt AG* were:

A direct starboard side view of the *P.1101* fighter version. Scale model by *Mike Hernandez*. Photographed by *Tom Trankle*.

• The cannon installation location set aside around the nose of the air intake duct was becoming very crowded as a result ammunition boxes and spent shell casing chutes were not working well and failing to give trouble-free operation.
• The *1101's* low thrust axis necessitated large changes in trim with every increase or reduction in engine thrust.
• The movement motion of the main landing wheel/gear upon retracting up into the side of the fuselage presented a serious enclosure problem and making a workable gear covering was becoming an engineering nightmare.
• The wing, turbojet engine, and landing gear loading were grouped in such as way that it was felt that an excessive number of "strong points" on the fuselage frame were being required.
• The aircraft was becoming heavier beyond its desired empty weight, with each new modification
and so anticipated flight performance 549 mph at sea level [884 km/h] and 608 mph at 23,000 feet [978 km/h at 7,010 meters] was much less than *RLM* minimum maximum speed specified in their request for bids.

RLM officials in Berlin were aware of the serious difficulties *Voigt* was having with his *Projekt 1101* over at *Messerschmitt AG-* Oberammergau and they expressed worries that it could not meet design specifications. Later the *RLM* announced that *Focke-Wulf's Ta 183* designed by *Hans Multhopp* as the winner. But since considerable work had been done on the *P.1101* as a possible production prototype, the *RLM* agreed to continue its financial support, although on a heavily reduced level and *Messerschmitt AG* was to finish it up as a flight test plane to prove out the swept back wing in flight.

Its highly doubtful that *Willy Messerschmitt* would have given up on his *Projekt 1101* had the *RLM* canceled their financial support. He had instructed *Voigt* to go ahead and complete the machine anyway. *Messerschmitt* returned to his original idea of using the *1101* as a *Mach 1* research vehicle with the primarily goal of evaluating the characteristics of highly swept back wings as powered by a single *HeS 011A* engine. Throughout the early mouths of 1945 work on the *1101* continued but at a very slow pace. Despite the desperate days facing war-torn Germany in early 1945, *Voigt* nonetheless was anticipating that the

1101's flight trials could start about mid June, 1945. It is not clear as to what turbojet unit would have powered the *1101* were it completed. The *HeS 011A* never did become available for field tests and the *1101* prototype held only an empty *011A* unit. *Voigt* would have had to settle for a single *Jumo 004B* with its 1,984 pounds [900 kilograms] of thrust. There are no details of the anticipated performance with the *004B*. Nonetheless, this all changed for the worse, of course, because American infantrymen arrived in the Oberammergau area about mid April, 1945.

Several weeks prior to the appearance of the Americans in Bavaria, W*illy Messerschmitt* had instructed his workers to microfilm the *Projekt 1101's* design drawings and hide the container away in the Oberammergau area. It didn't matter much because American intelligence had the *1101* as a war prize and then proceeded to turn it into a piece of playground equipment. The French eventually found the canister's burial site and dug up the plans and took them back to Paris. As a typical gesture of thanks for the liberation of France mostly by American infantry, French intelligence refused to share the *1101's* plans with anyone even though the Americans had asked for a copy from time to time.

Had it not been for the increasing weight problem, the P*rojekt 1101* would have had several advantages. For example it appears to have been a relatively easy machine to maintain and service. Also, *Voigt's* high altitude fighter design called for its wings to swept back at 40° at quarter chord. The two-piece wing with a root thickness of 8% (measured parallel to plane of symmetry) with a tip thickness of 12% would have been mounted shoulder high and constructed with a steel main spar with wooden ribs and plywood covering. It would have had full-span automatic leading-edge slots, plain chamber-changing trailing-flaps, and conventional inset-hinge ailerons, elevator, and rudder controls.

The wing tips were made relatively thick because (a) this thickness could be tolerated without compressibility difficulty in view of the three dimensional flow at the tip (b) greatest attainable torsional rigidity was desired to maintain aileron control at a high speeds (c) the thicker wing per se and the larger chord slat that could be fitted to it would both help to prevent tip stall. An incidental advantage was that both the slat and the flap could be made of constant cross section throughout their span. The resulting slat chord was 13% of the wing chord at the root and 24% at the tip.

The empennage (tail unit) was conventional except that it had rather low aspect ratio, high taper, and large sweep back of the leading edges of the fin and stabilizers. No armament appears in the prototype since it was no longer viewed as a fighter but only as a research machine. The *1101's* barrel-like fuselage had a slender tailboom (so characteristic of *Messerschmitt* military and civilian designs) aft of the jet pipe, a pressurized cockpit set well forward, and a direct nose air intake. The single small diameter nose wheel was to retract with a 90° turn up into the space beneath the air intake and the main undercarriage wheels, rearward, and inward by means of a skewed axis joint at the upper oleo fitting. The main landing gear was pivoted at the fuselage, just under the wing and retracted aft.

The pressurized cockpit was located above the duct and forward end of the engine, making for a rather deep fuselage. The jet discharge was located in a half-tunnel step near the root trailing edge. The aircraft was a mid wing, with the wing structure passing under the fuel tanks but above the He*S 011A* engine.

Jet fuel for the single *HeS 011A* was carried in three individual fuel tanks inside the fuselage above the engine in a line aft the pilot's seat back. Total carrying capacity is unclear but it has been stated of ranging between 345 and 370 gallons. Armor plating would give the pilot protection against 12.7 mm cannon fire forward and 22 mm cannon fire from the rear. Offensive cannon included three *MK 108*...two on one side the air intake duct and one cannon on the other. Service ceiling was estimated at 46,000 feet [14,021 meters]. Were the *1101* to have been fitted with a *Jumo 004B*, the aircraft would not have reached that altitude.

A camouflaged *P.1101* seen from its rear starboard side. Shown is the complicated main wheel gear cover door and its single round jet pipe coming from the 2,866 pound thrust *HeS 011A* axial turbojet engine. Scale model by *Mike Hernandez*. Photographed by *Tom Trankle*.

Immediately post war American military intelligence officers and their civilian counterparts were eager to speak directly with Voigt about the Projekt 1101 as well as other proposed company-wide Messerschmitt AG turbojet-powered fighters and bombers. They had their opportunity in July, 1945 and Americans spoke at length with Voigt. He was highly articulate with a good command of the English language. It was during these interrogation sessions that American officials learned the dirty little truth about the ugly Projekt 1101, a full litany of its design faults, and how the prototype fighter/research aircraft seemed to defy all attempts of the best minds within Messerschmitt AG to mold it into a workable machine. With this knowledge learned, American intelligence interest in the 1101 virtually ended and Willy Messerschmitt's pride and joy was pulled out of its workshop at Oberammergau and placed on static display. The engineless Me 262 found, too, next to the 1101 was also placed out on the grass. Out there on the grass the 1101 probably would have remained indefinitely but except for its discovery by the late Robert Woods. He was a member of the Army's civilian intelligence evaluating team. He was Bell Aircraft Corporation's chief designer. Bell had first seen the 1101 out on its grassy display while visiting American intelligence headquarters which had been established at Oberammergau immediately post war. He told the American intelligence people there that if all the 1101 was going to be used for static display then he'd like to study it more closely back at Bell Aircraft Corporation facilities in Buffalo, New York. The United States Army complied with Woods' request given him the experimental Messerschmitt aircraft. Its wings were removed and the badly deteriorating aircraft crated up and transported to Bell Aircraft in Buffalo. While serving static display duties the 1101's steel wing spars were rusting, the wooden wing covering was coming loose at the joints, its starboard main landing gear oleo strut had collapsed, and its fuselage tail boom had

been bent as the aircraft sunk into the earth. Nevertheless the American military transported the 1101 as it was to Wright-Patterson Air Force Base (AFB) outside the city of Dayton, Ohio. Woods also enticed Voigt into coming to America and Bell Aircraft to help explain the 1101's variable wing geometry to the Bell aviation designers. Thus in 1946 under "Operation Paperclip," Voigt and his family entered the United States as a desirable German alien as did several hundred other former German military equipment designers, scientist, and inventors. In the Autumn of 1946 the P.1101 arrived at Wright-Patterson AFB from Oberammergau after serving as an outdoor static display for about one and half years. The machine was not a pretty sight. Furthermore, it remained in outdoor storage at Wright-Patterson AFB until 1948 when it was transported to Bell Aircraft at Buffalo. Ultimately it would become a nonflying full scale mockup at Bell for two identical turbojet-powered experimental aircraft built by Bell and known as the X-5. One of the X-5s made its maiden flight on 20 June 1951, and the machine became Bell's official offer to the U.S. Air Force as a high-speed and high altitude interceptor. The Air Force rejected the machine after a series of exhaustive flight tests. The sole surviving prototype was retired in October, 1955.

If Projekt 1101 did not create any problems for the Allies during the war, it certainly was the cause of a number personal confrontations in America, particularly within Bell Aircraft. Trouble began almost immediately when the 1101 arrived at Bell and Robert Woods announced to the design staff his ideas concerning the former Messerschmitt AG machine. Woods had the ear of Larry Bell, owner and president of Bell Aircraft. Woods had persuaded Bell that the P.1101 was a design and engineering feat so great that its potential should not be ignored. Larry Bell assigned Robert Stanley, the company's chief engineer, to study the beat-up rusting hulk formerly known as Messerschmitt Projekt 1101 with the thought of building one like it as a salable military fighter/interceptor aircraft.

Robert Stanley put his best staff engineers on the 1101 project evaluation. After a thorough review, Stanley and his staff reported to Larry Bell, Robert Woods, and others in a management meeting that the 1101 was currently inferior to existing American turbojet-powered fighter/interceptors already in the air, the F-86 for example. Robert Woods was outraged. How could they not see the genius in this design? Woods rejected the Stanley team's conclusion and continued to champion the merits of the 1101 finally convincing Larry Bell to proceed with the construction

Direct port side view of a 1101 armed with 2x X-4 guided air-to-air missiles. Scale model by Dan Johnson.

of two prototypes called the *X-5*. *Larry Bell's* decision to proceed with the *X-5* prototype based on the *1101* created a great division within the upper ranks of *Bell Aircraft*. Feelings of the *Stanley* review team against the Americanization of the *1101* were so strong that several of his engineers resigned. *Stanley*, too, resigned his position as chief engineer with *Bell Aircraft*. It made little difference with *Bell's* chief designer *Robert Woods*. Unswayed, actually feeling good riddance regarding the several key defections of good men, *Woods* took over *Stanley's* former position as chief engineer and saw through the Americanization of *Projekt 1101*.

Robert Stanley and the others who resigned from *Bell Aircraft* over *Larry Bell's* decision to built the *X-5* based on *the1101* later learned that their evaluation and calculations had been correct. Most of their fears that the Americanization of the *1101* was a mistake because it would lack endurance, speed, range, and weapons carrying ability were all confirmed in flight testing by the U.S. Air Force. In addition, the design was considered inappropriate for the aircraft's purpose, that is, as an interceptor because of its small size and powered by only a single *HeS 011A* turbojet engine. In adequate power for any type of interceptor. The U.S. Air Force stated that the "swing-wing" was su-

perfluous option on an interceptor intended to serve as a last line of defense...in the *1101's* case as a interceptor for high flying *B-29* "*Superfortress*" bombers. Furthermore, weapons on the "swing- wing" fighter had to be carried almost entirely inside the fuselage, as the *General Dynamics F-111* fighter later demonstrated, and extra wing loading was nearly impossible. Indeed, although the U.S. Navy believed then that a ship-based interceptor might be needed, the U.S. Air Force did not need an aircraft with home interception capabilities during the 1950s.

Few people, then nor now, consider the M*esserschmitt P.1101* very pleasing aerodynamically, your author to the contrary. However, the *1101* generally receives high marks for *Voigt's* inventiveness in terms of system's layout, state-of-the-art engineering, and airframe design. The simple layout of the turbojet engine would have made maintenance and engine change out easy for *Luftwaffe* field- based mechanics. Thus, field servicing could have been accomplished quickly. The *1101* might have served Germany's needs nicely, especially in late 1944 and early 1945 when *B-17s* and *B-25s* were looking freely around the countryside seeking to hit anything of value. But whether it could have made a real difference in Germany's dying war-effort is not due to its limits as a gun platform.

The sole example of the M*esserschmitt/Voigt P.1101* was junked at *Bell Aircraft* in the early 1950s after it was used for structural tests and competition of the two *Bell X-5's*. Scraping the *1101* was unfortunate and unnecessary like the burying of several *Arado Ar 234Bs* discards at the end of the U.S. Navy's Patuxent River Naval Air Station (NAS), Maryland runway in the mid1960s. But this was the late1950s to mid 1960s and few people cared about keeping these ex-*Luftwaffe* machines of war for future generations to see and touch. Two *X-5s* for built but only one *Bell X-5* survives. It was the world's first airplane to vary the sweep back of its wings in flight. Built to prove the theory that by increasing the sweep back of an airplane's after takeoff, a higher maximum speed could be obtained, while still retaining a relatively low take-off and landing speed and higher rate of climb with the wings swept forward. The first *X-5* flight was made on 20 June 1951 by "*Skip*" *Ziegler* at Edwards Air Force Base, California. On the machine's 5th or 9th flight (it is not entirely clear), its wings were operated through the full (1st prototype) sweep range of 20° to 60°. One of the two *X-5s* crashed and was destroyed on 13 October 1953 when it failed to recover from a spin at 60° sweep back. The surviving *X-5* was retired in October, 1955 and delivered to the U.S. Air Force Museum, Wright Patterson, AFB, Dayton, Ohio in March, 1958.

Messerschmitt P.1101 Specifications:

- Type: Fighter...later research aircraft
- Country: Germany
- Manufacturer: Mes*serschmitt AG*
- Designer: W*oldemar Voight*
- Year Constructed: 1944
- Power Plant: 1xHe*S 011A* producing 2,866 pounds [1,300 kilograms] of static thrust
- Wing Span: 27 feet 1 inch [8.2 meters]
- Wing Area: 171 feet squared [15.9 meters squared]
- Wing Loading: Na
- Aspect Ratio: Na
- Length: 29 feet [9.1 meters]
- Height: 9 feet 2 1/2 inches [2.8 meters]
- Weight, Empty: 5,719 pounds [2,595 kilograms]
- Weight, Takeoff: 8,966 pounds [4,067 kilograms]
- Weight, Maximum Flying: Na
- Fuel, Internal: about 370 gallons
- Crew: 1
- Speed, Maximum: 550 mph at sea level [885 km/h] and 610 mph [985 km/h] at 22,960 feet [7,000 meters]
- Speed, Cruising: Na
- Speed, Landing: 107 mph [172 km/h]
- Takeoff Distance: 2,361 feet [720 meters]
- Range, Maximum: 932 miles [1,500 kilometers]
- Flight Duration: 40 minutes
- Ceiling: 45,931 feet [14,000 meters]
- Rate of Climb: 4,370 feet/minute [1,520 meters/minute]
- Armament: None
- Bomb Load: None
- Status: Scrapped at Bell Aircraft, Buffalo, New York (early 1950's)

Bell Aircraft X-5 Specifications:

- Type: Research aircraft to vary the sweep back of its wings in flight...first American fighter aircraft to do so
- Country: USA
- Manufacturer: Bell Aircraft
- Designer: Group design based on the *Messerschmitt AG P.1101* of 1944
- Year Constructed: 1950/1951
- Number Constructed: 2
- Maiden Flight: 20 June 1951
- Power Plant: Alli*son J35* A-17 producing 4,900 pounds [2,223 kilograms] of static thrust
- Wing Span, Wings Extended: 32 feet 9 inches [10.3 meters]
- Wing Span, Wings Swept: 22 feet 8 inches [6.0 meters]
- Wing Area: Na
- Wing Loading: Na
- Aspect Ratio: Na
- Length: 33 feet 4 inches [10.2 meters]
- Height: 12 feet [3.6 meters]
- Weight, Empty: Na
- Weight, Take Off: 9,892 pounds [44,870 kilograms]
- Weight, Maximum Flying: Na
- Fuel, Internal: Na
- Crew: 1
- Speed, Maximum: 690 mph [1,110 km/h]
- Speed, Cruise: 600 mph [966 km/h]
- Speed, Landing: Na
- Take Off Distance: Na
- Range, Maximum: 500 miles
- Flight Duration: Na
- Ceiling: 50,700 feet [15,453 meters]
- Rate of Climb: Na
- Armament: None
- Bomb Load: None
- Status: 2nd prototype crashed/destroyed and 1st prototype on permanent outdoor static display at the U.S. Air Force Museum, Wright-Patterson AFB, Dayton, Ohio.

Opposite: Direct starboard side view of a *1101* showing its mid-fuselage mounted swept back wing. Scale model by *Dan Johnson*. *Right:*An overhead port side view of the *P.1101* in camouflage. The *1101's* wing sweep back was set at 40 degrees. Scale model by *Mike Hernandez*. Photographed by *Tom Trankle*.

Oberst Siegfried Knemeyer, Chef/TLR (Chief of Technical Air Armament) at the RLM and the powerful man who was behind the so-called "Emergency Fighter Competition" set forth in late 1944. Knemeyer issued specifications to all the principal aircraft companies requesting bids for a new single-engine HeS 011A-powered fighter. These fighters were to have a 621 mph [1,000 km/h] level speed at 22,960 feet [7,000 meters] and have the ability to operate comfortably at altitudes up to 45,920 feet [14,000 meters]. By February, 1945 Knemeyer had received a number of responses: two from Focke-Wulf, three from Messerschmitt AG, one from Heinkel, one from Junkers, and one from Blohm und Voss. Knemeyer is shown leaving the cockpit of the motorized Horten Ho 3G sailplane which he had just flown. He was fond of Horten all-wing machines and would later instruct Reimar and Walter Horten to construct the four HeS 011A turbine engine-powered Ho 18B "Amerika Bomber" in early 1945. Oranienburg. About Autumn, 1944.

This is the demon Siegfried Knemeyer's Luftwaffe pilots had to go up and conquer or Germany would certainly die...the dreaded American Air Force's B-29 "Superfortress." Officials at the RLM anticipated this heavy bomber over Germany any day by mid 1944. However, it was never assigned bombing duties over Nazi Germany. The B-29 had sufficient power to comfortably reach altitudes of 35,000 feet because of its four 3,342 cubic inch Wright R-3350 Duplex Cyclone 18-cylinder two- row radial engines. Each one was fitted with two General Electric turbo-superchargers and the 2,670 pound engine provided 2,200 horsepower at 2,800 rpm. The B-29 had a range of 3,250 miles and a bomb load of 16,000 pounds verses a bomb load of 6,000 pounds for a B-17G.

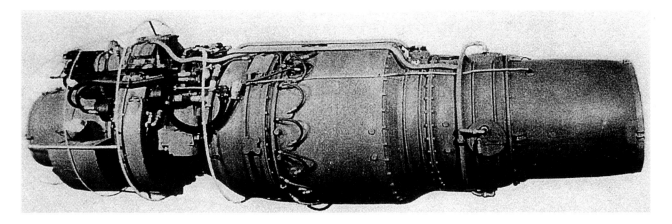

A port side view of the *Heinkel-Hirth HeS 011A* turbojet engine with a design rating of 2,866 pounds
[1,300 kilograms] static thrust. Its air intake is left
with the jet pipe to the right. A single *HeS 011A* was
to have powered all the entries in *Oberst Siegfried
Knemeyer's* "*Emergency Fighter Competitions*" of
late 1944.

A nose starboard side view of a *HeS 011A* turbojet
engine. The air intake is to the right and showing the
first stage compressor fan.

Dipl.Ing. Woldemar Voigt, design genius of *Messerschmitt AG* and shown with eyes closed ...perhaps lost in thought or just taking a "cat nap." About mid April, 1939...the time he had completed the design work on the *Me 262.*

The legendry *Willy Messerschmitt.* This photo is from the mid 1940s.

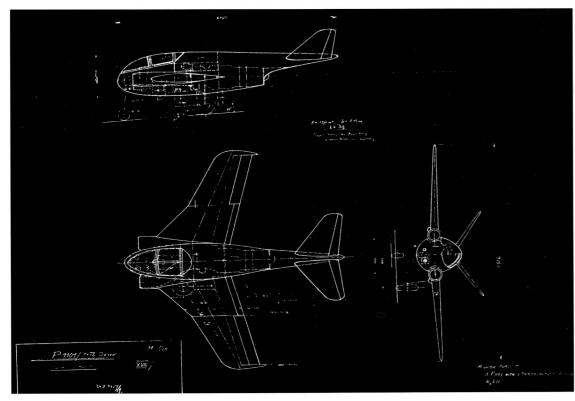

This the first known design for *Messerschmitt Projekt 1101*. It dates from July, 1944. It was to have a large "greenhouse-type of cockpit with plenty of plexiglass. Circular air intakes for the turbojet engine were located at the wing root port and starboard and this project design is unusual also in that the *Messerschmitt* designers wished to give it a "*butterfly*" or "V" tail...very unconventional at the time although *Messerschmitt AG* had collected practical data on this configuration through the modification of a *Bf 109F*.

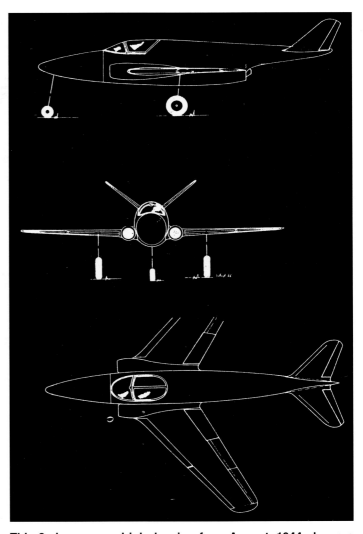

This 3-view pen and ink drawing from August, 1944 shows a further refined *Projekt 1101* with the "V" tail but with a pointed nose and far less plexiglass about the cockpit which beginning to look more and more like the final *1101* concept.

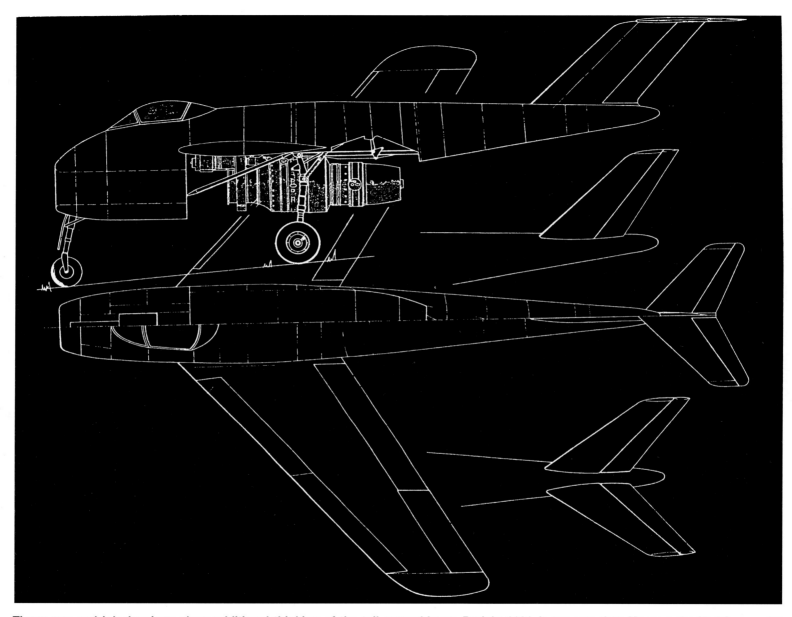

These pen and ink drawings show additional thinking of the tail assembly on *Projekt 1101.* It appears that *Messerschmitt AG* was still undecided as to the best tail assembly on what seems to be a mature *1101* fuselage, however, it can be seen with a *Hans Multhopp* T-tail and the *Messerschmitt AG* "V" shaped tail unit.

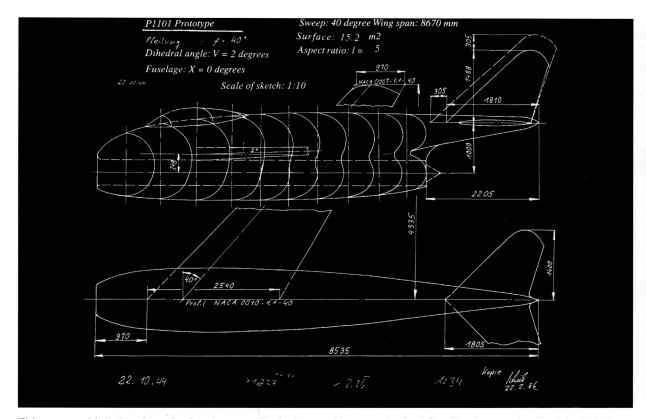

P1101 Prototype

Pfeilung : φ = 40°

Dihedral angle: V = 2 degrees

Fuselage: X = 0 degrees

22.10.44

Scale of sketch: 1:10

Sweep: 40 degree Wing span: 8670 mm

Surface: 15.2 m2

Aspect ratio: l = 5

NACA 0007-1,1-40

Profil NACA 0010-1,1-40

22.10.44

Kopie 22.2.86

This pen and ink drawing of a fuselage profile is from a *Messerschmitt AG* collection on the *Projekt 1101* from October, 1944 and is believed to have been used for a wind tunnel model.

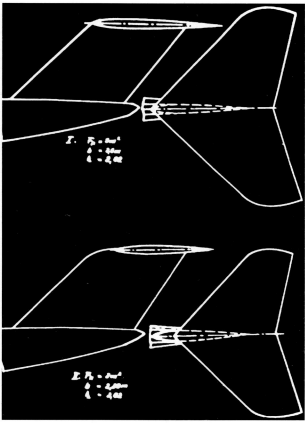

The "T-tail" proposed for the *1101* was classic *Hans Multhopp* T-tail. Shown are two designs for the horizontal stabilizer on top of the vertical fin. The upper drawings shows the stabilizer's apex to be just above the vertical fin's leading edge while the lower drawing shows the horizontal stabilizer moved aft.

Woldemar Voigt shown here in the early 1940s. He wore this hair style pretty much his entire life.

A highly detailed port side view of Voigt's P.1101 as it would have looked when completed about June/July, 1945. Scale model by **Günter Sengfelder**.

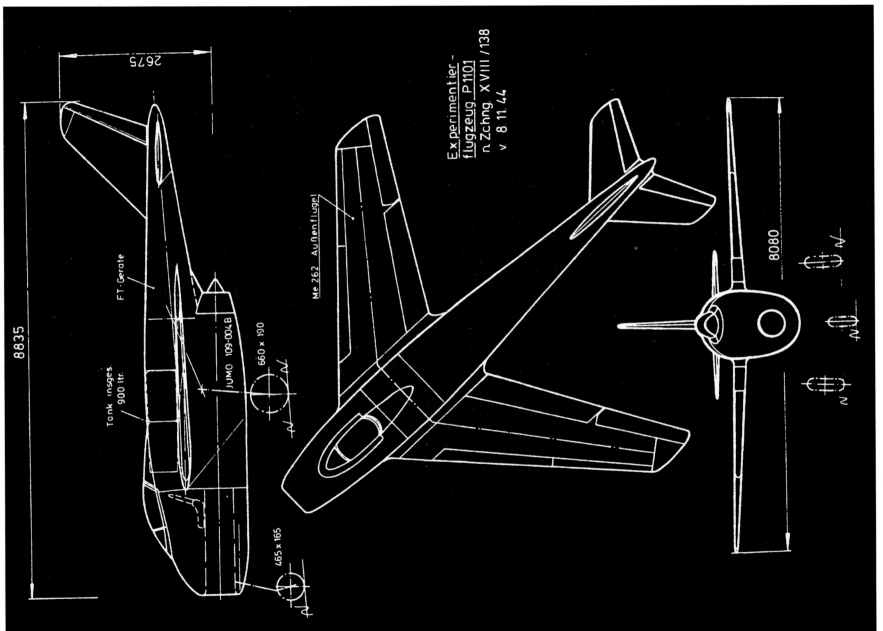

Experimentier-
flugzeug P1101
n Zchng. XVIII /138
v 8.11.44

2675

8835

FT-Gerate

Tank insges.
900 ltr.

JUMO 109-004 B

660 x 190

465 x 165

Me 262 Aussenflugel

8080

A pen and ink drawing of the "Experimental" *Projekt 1101* as it came to be called in August, 1944. This was after the *1101* had lost out in design competitions against several others...the *Fw Ta 183* by *Hans Multhopp* being the winner. *Messerschmitt* decided after losing the competition to complete the machine and use it for flight research.

A port side view of the P.*1101* painted in a silver color and as it probably would have looked when taken out side after its anticipated completion about June/July, 1945. Shown here with no armament, national markings, and so on. Scale model by *Günter Sengfelder*.

An air intelligence photograph of the entire *Messerschmitt AG* complex at Oberammergau in southern Bavaria. With walls surrounding the complex it took on the appearance of a walled mediaeval city.

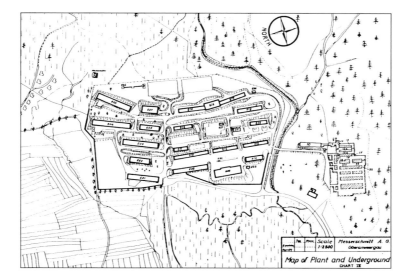

A pen and ink drawn map by the U.S. military intelligence of the former *Messerschmitt AG* complex including underground facilities of Oberammergau. About mid 1945.

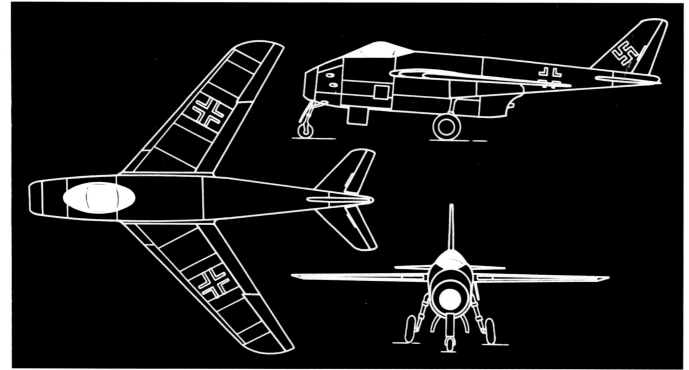

The final design layout of *Voigt's Projekt 1101.* Notice that is shows no military equipment or armament has been drawn on the *1101.*

A view of the former *Messerschmitt AG* complex at Oberammergau as viewed from a hill-side south and east of the walled-in facility/city.

A close-up view of the former M*esserschmitt AG* complex at Oberammergau design offices from a hill side high above the walled-in facility/city.

A direct overhead view of the P.*1101* showing its 40° wing and tail plane sweep back. Scale model by *Dan Johnson*.

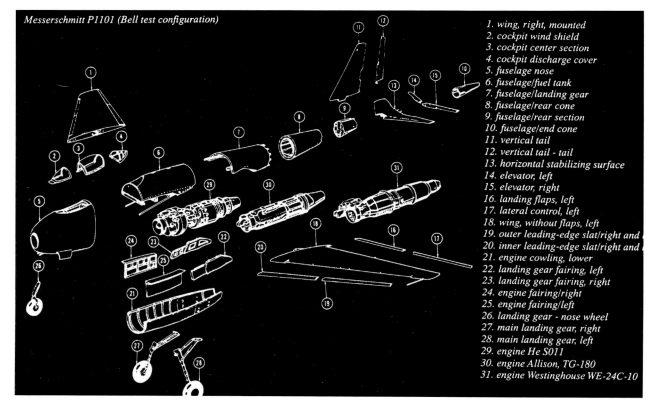

Messerschmitt P1101 (Bell test configuration)

1. wing, right, mounted
2. cockpit wind shield
3. cockpit center section
4. cockpit discharge cover
5. fuselage nose
6. fuselage/fuel tank
7. fuselage/landing gear
8. fuselage/rear cone
9. fuselage/rear section
10. fuselage/end cone
11. vertical tail
12. vertical tail - tail
13. horizontal stabilizing surface
14. elevator, left
15. elevator, right
16. landing flaps, left
17. lateral control, left
18. wing, without flaps, left
19. outer leading-edge slat/right and
20. inner leading-edge slat/right and
21. engine cowling, lower
22. landing gear fairing, left
23. landing gear fairing, right
24. engine fairing/right
25. engine fairing/left
26. landing gear - nose wheel
27. main landing gear, right
28. main landing gear, left
29. engine He S011
30. engine Allison, TG-180
31. engine Westinghouse WE-24C-10

The *Projekt 1101* with its main components disassembled. Courtesy *Aerofax/Jay Miller*.

A port side view of the *P.1101* looking about the way the American infantrymen found it at the abandoned *Messerschmitt AG*-Oberammergau facilities.

A direct nose-on view of the restored *P.1101* and shown at *Bell Aircraft*, Buffalo, New York about Summer, 1950.

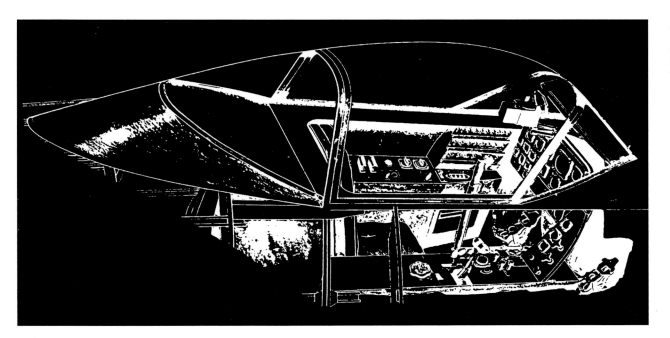

A pen and ink drawing of the so-called *Messerschmitt* "racing cockpit" canopy to be used on the *Projekt 1101.* It was the same type of canopy used on the production *Me 262* and it allowed for a full all-around view.

A close up view of the *Messerschmitt* "racing" cockpit canopy installed on the *P.1101.*

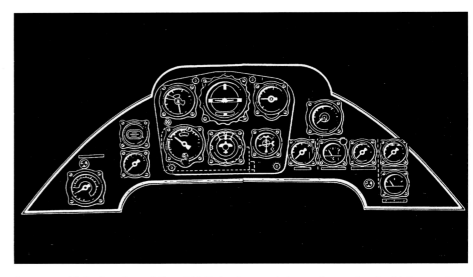

A pen and ink drawing of the *1101's* instrumental panel...much of which was borrowed from the *Me 262.*

The *P.1101* used the typical *Messerschmitt* side-hinged cockpit canopy. It opened to the starboard side as seen in this photo of the *1101* at Oberammergau before it the machine was removed outside to assume static display duties.

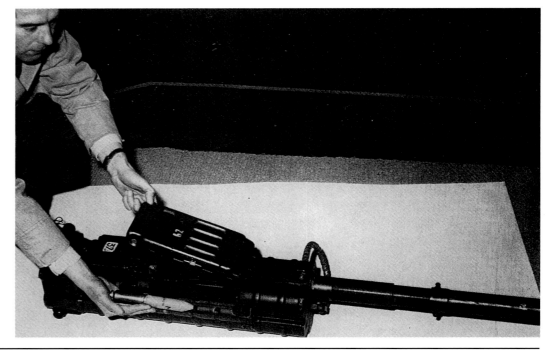

The *1101* was to have been armed with *MK 108* cannon. Shown in this photo is highly regarded aviation historian *Alfred Price*, postwar with a *108* of the type which would have been placed in the military version of the *1101* had it been ordered into serial production by *Oberst Siegfried Knemeyer*.

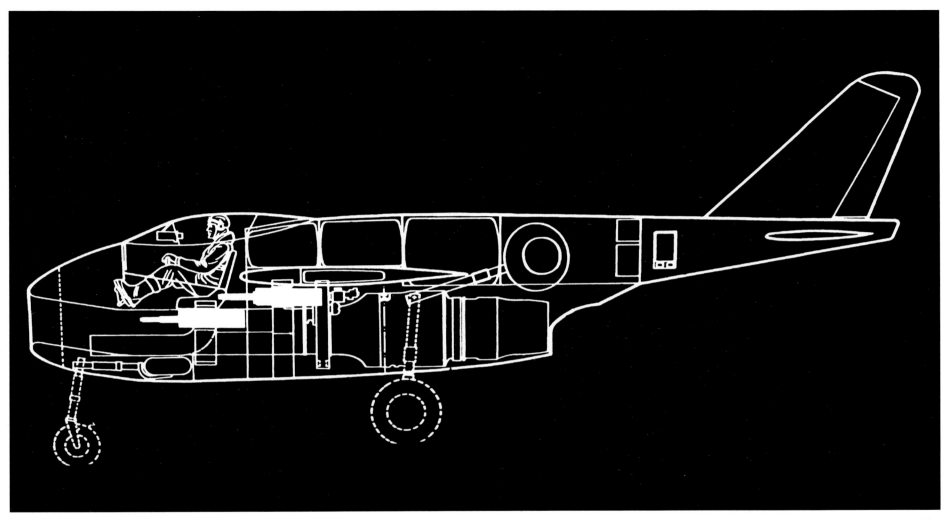

A port side view of the military version of the *1101*. Up to *4xMK 108* 30 mm cannon would have been preferred on the *1101...two* mounted each side of the fuselage. The square box just forward of the vertical fin's base is the aircraft's compass.

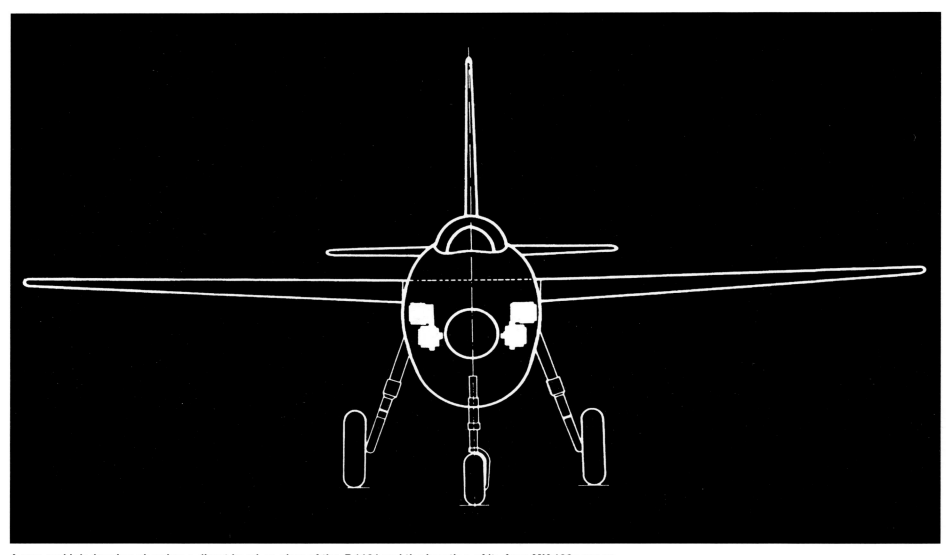

A pen and ink drawing showing a direct head on view of the *P.1101* and the location of its four *MK 108* cannon.

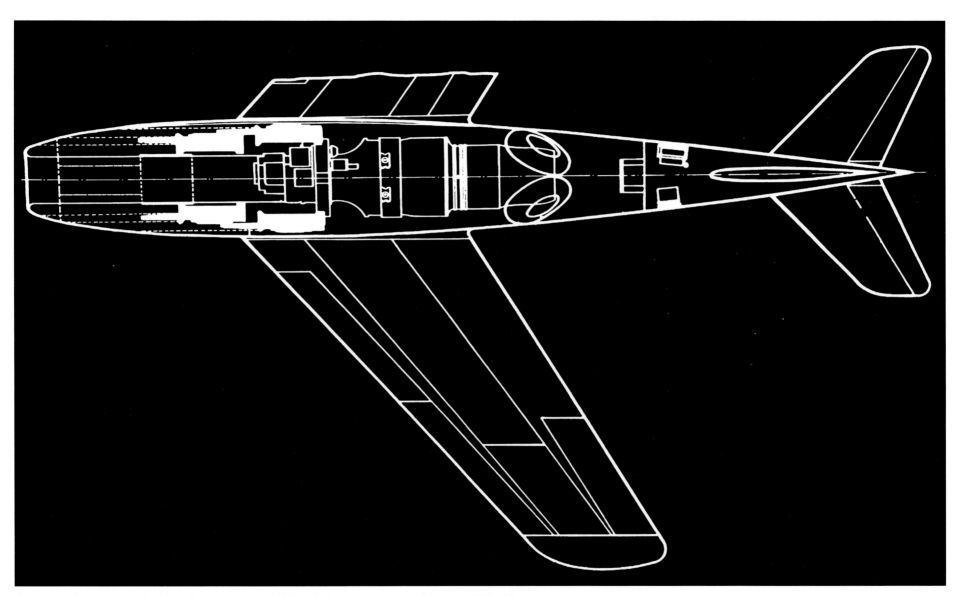

A pen and ink drawing looking down on the *P.1101* and the arrangement of its four *MK 108* cannon.

The P.1101's fuselage narrowed considerably as it approached the tail assembly. A narrow aft fuselage was a typical *Messerschmitt AG* design characteristic.

The *Projekt 1101* with its *HeS 011A* uncovered as it appeared to American infantrymen in April, 1945, although intelligence reports state that the machine did not have its wings mounted. Scale model by *Günter Sengfelder*.

This is a photo taken by American infantrymen of the *P.1101* as it was found. To the right is a *Me 262* without engines installed and it has fallen back on its tail. Photos of this facility show it to be very messy—although it was never bombed—and the *1101's* nose and port tire nearly flat.

A rear starboard side view of the *P.1101* at Oberammergau. The tail cone has been installed making for a nice rounded end for the cone-shaped fuselage. It appears that the *1101* is being prevented from falling back on its tail through the use of a hydraulic lift...its arm which is seen at the bottom of the photo.

Germany's 2nd generation turbojet engine—the *HeS 011A* axial flow turbojet rated at 2,866 pounds [1,300 kilograms] static thrust. The *Luftwaffe* had great hopes of glory from this engine and it was to have powered numerous heavy fighters as well as bombers. But it was not ready for field testing when the war ended in Europe in May, 1945. It may have taken another twelve months to perfect.

A full starboard side view of the *Projekt 1101* as found in April, 1945. Shortly after this photo was taken, American intelligence painted "*Me - 1101 V1*" on its nose. Then shortly afterward it was moved outdoors to become a static display for nearly two years.

A starboard side view of *Projekt 1101's HeS 011A* installed as found. It is not clear whether this engine was new and ready to run or merely an empty shell of the 2,866 pound thrust engine.

A close up view of *Projekt 1101's HeS 011A* engine installation as seen from the starboard side. *Voigt's* design allowed ample room for placing turbojet engines from different manufacturers in the machine, in fact, post war the American-made turbojet *Allison J35 A-17* with its 4,900 pound [2,223 kilograms] static thrust fit the air frame very nicely, too. Scale model by *Günter Sengfelder*.

A camouflaged *1101* as seen from its rear port side. What is interesting is the clear way the jet pipe separates from the fuselage leaving the *HeS 011A's* exhaust unhindered. Scale model by *Mike Hernandez*. Photographed by *Tom Trankle*.

A full rear view of a *1101* showing how the *HeS 011A's* jet pipe exits nicely under its tail boom. Scale model by *Dan Johnson*.

A port side view of a *1101* which is carrying the air-to-air *X-4* guided missile. Scale model by *Dan Johnson*.

A nose view of a *1101* from a few feet above the aircraft featuring its four *X-4* wing-mounted air-to-air guided missiles. The nose wheel/gear cover is not correct...it is about half as wide as it should be but otherwise a very nice job. Scale model by *Dan Johnson*.

A direct nose-on view of a *1101* featuring its shoulder-high wings with their nearly uniform thickness from wing root to wing tip. Scale model by *Dan Johnson*.

A *1101* appearing in a forest setting just as if it had been assembled and complete except for national markings. The nose wheel/gear cover size is correct. Scale model by *Günter Sengfelder*.

Woldemar Voigt's Projekt 1101 in the United States at *Bell Aircraft*, Buffalo, New York in the late 1940s. It has been restored and still carries its *HeS 011A*, however, the machine carries templates showing where three of its six cannons would be located on *Bell's* version of *Voigt's 1101* being called the *X-5*.

A rear view of a serial production *1101* featuring its fixed 40° swept back wings. Scale model by *Dan Johnson*.

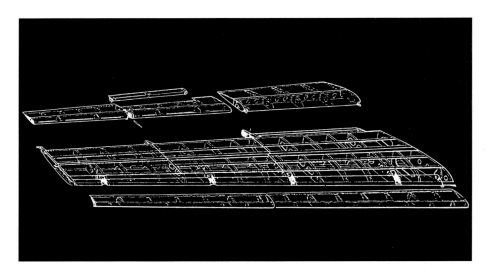

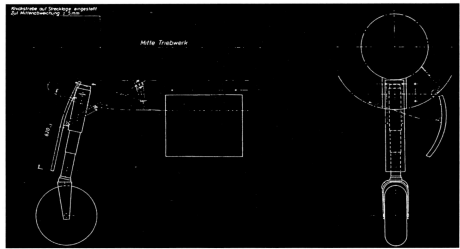

Above: A pen and ink drawing of a *1101's* starboard wing internal structure. The wing was similar to that of a *Me 262* from about rib 7 through the final rib [rib #21]. The *1101's* wing profile, aileron section and leading edge flaps was similar to those found in the *262*, too, however its leading edge flaps had been extended over almost the entire wing span were enlarged to 20% from 13% as found on the *262*. The *1101's* wing was attached to the fuselage at three points.

Above Right: A pen and ink illustration of the *Projekt 1101's* nose wheel assembly including its long narrow door cover...from the port side and front-on.

The small, simply-built nose wheel assembly the *1101* as seen from its starboard side. Nose tire size was about 18.3 inches by 6.5 inches, or 465x165 mm in diameter.

Voigt's 1101 at *Bell Aircraft* featuring its nose wheel assembly, however, minus its door cover. The square access port seen at the far right is gain access to the *HeS 011A's Riedel* turbine starter motor. A similar access port was located on the opposite side of the fuselage.

A *1101* nose section featuring its cockpit canopy and its nose wheel/gear assembly with its two-piece gear cover doors. Scale model by *Günter Sengfelder*.

The *1101* at *Bell Aircraft.* The height of the port wing is apparent from the engineer standing beneath. The size of the main wheel oleo strut is significant, too. The *1101* still carries its *HeS 011A* turbojet engine. Later it would be removed and a *Allison J35* installed to determine fit and alinement. A complete *HeS 011A* was a very rare engine at the end of the war. Were it still around it would be the rarest of the rare in any aviation museum.

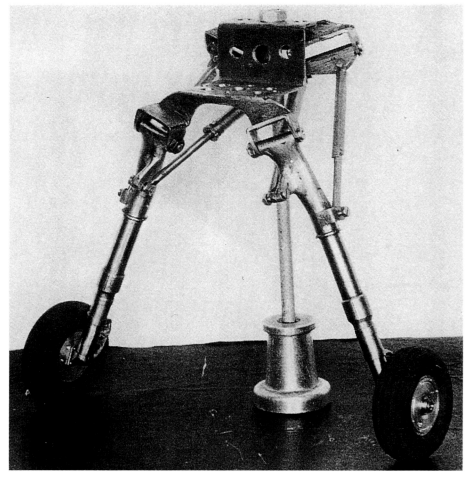

The main wheel landing gear assembly for the *Projekt 1101* fully extended . Only the main wheels had brakes. Scale model by *Günter Sengfelder*.

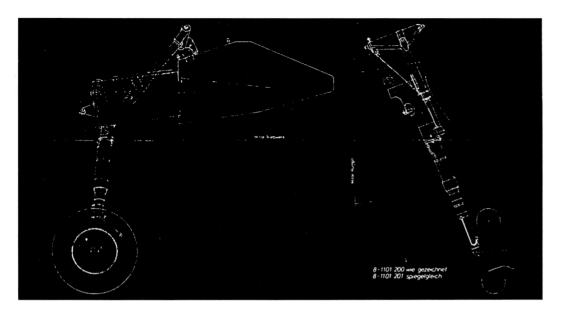

A pen and ink drawing of the main landing gear for a *1101*. To the left is the port side view and on the right is also a head-on view of the landing gear assembly.

A starboard side view of the *1101* prototype at *Bell Aircraft* with a *Allison J35* turbojet engine installed in place of the *HeS 011A*.

The main landing gear assembly of the *1101* fully retracted. The main tire size on the prototype was 660x190 mm but serial production would have used a 29.1 inch by 8.3 inch, or 740x210 mm diameter tire, instead. Scale model by *Günter Sengfelder*.

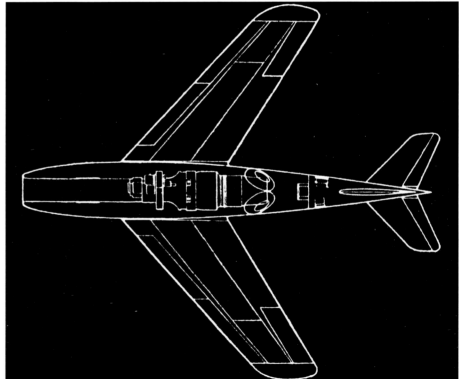

A pen and ink drawing of the *Projekt 1101* showing the approximate location for the fully retracted main landing gear wheels near the end of the *HeS 011A's* jet pipe. Notice, too, the long air intake duct going back to the engine's nose and there was considerable concern that the long duct would rob the turbine of much of its power.

A series production *1101* as seen ground level from its nose starboard side and armed only with 4x*MK 108* cannon. Digital image by *Andreas Otte*.

Four of the five proposed fighters entered in *Siegfried Knemeyer's* "*Emergency Fighter Program Competitions*" in mid 1944. Left to right: *Blohm und Voss' 215*, *Junkers EF 128*, *Focke-Wulf's183*, and in front the *Messerschmitt AG's Projekt 1101*. Missing from this photo is *Heinkel AG's* bid and known as *P.1078A*. Scale models by *Dan Johnson*.

Woldemar Voigt at his home in Annapolis, Maryland prior to his death in June, 1980.

The *Focke-Wulf Ta 183* by *Hans Multhopp* and the winner of *Knemeyer's Emergency Fighter Program Competitions.*

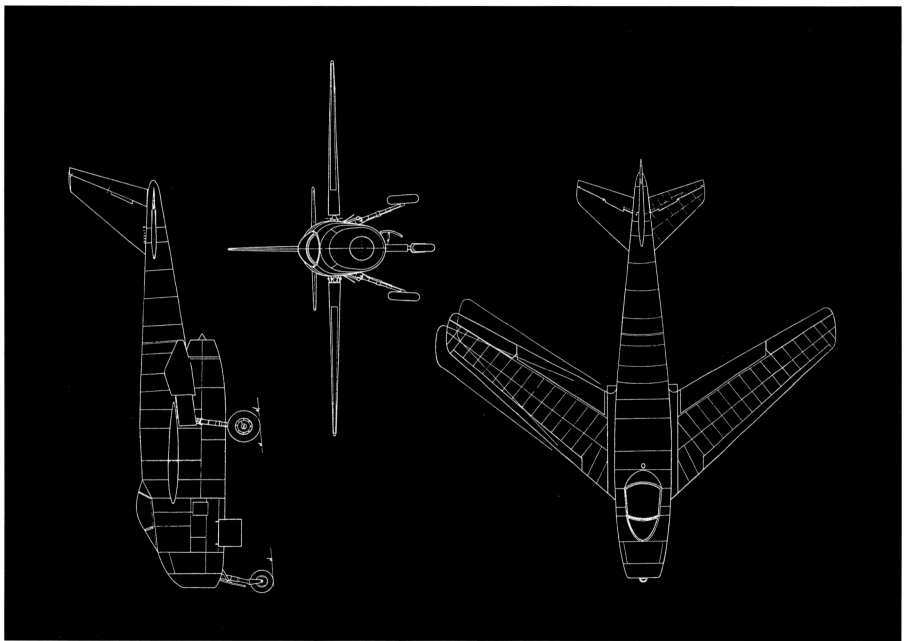

A pen and ink three-view drawing of serial production version of the *Projekt 1101* based on original documentation. The full overhead view shows the wings with two positions of sweep back: 35°s and 40°s.

An overhead view of a silver painted *1101* with its swept back wings fully extended for take off and landing at 35°s. Scale model by *Ed Bailey*.

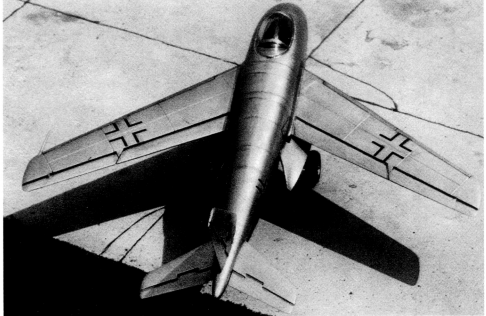

In this photo the *1101's* swept back wings have been moved aft to 40°s of sweep back. Scale model by *Ed Bailey*.

Two members of the American Army Air Force Intelligence team (right) plus driver (left) photographed at Oberammergau about mid May, 1945.

A poor quality photo of the port side view of the *1101* shortly after the Americans arrived at Oberammergau in April, 1945. The tail of the *1101* is being supported by a hydraulic lift to keep the aircraft from falling back on its fragile tail section.

To the right in the photo is the *1101* its tail being supported by a jack-stand. In the far right of the photo is a *Me 262* minus its *004B* turbojet engines. Without engines to *262* to keep it in balance it has fallen back on its tail.

A poor quality photo of the *1101* as seen from the rear starboard side at Oberammergau in April, 1945. The workshop where the machine was discovered by the U.S. Army was a mess. It has been reported in U.S. Military Intelligence documents that the *1101* was found without its wings attached. This might account why in photos of this period the *1101* tail is supported by a jack-stand. In this photo American Military Intelligence have taken the jack-stand away and placed a hydraulic long-arm lift which can be seen in lower center of the photo. This author is unaware of any photos showing the 1*101* without its wings attached.

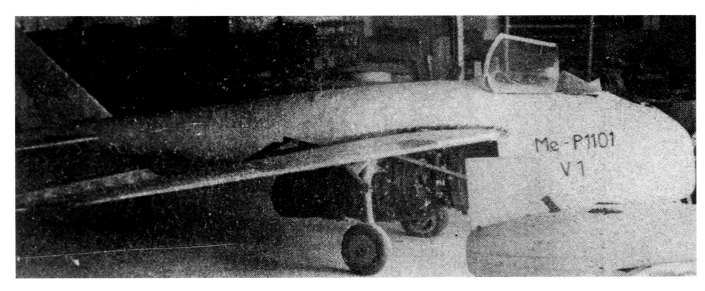

A poor quality photo of the *1101* at Oberammergau with its typical *Messerschmitt-style* cockpit hatch open. A member of the U.S. Army has painted "*Me - P1101 V1*" on the *1101's* starboard side forward nose. To the far right of the photo is the outline of a Messerchmitt AG-designed *E-4* flying bomb.

A close-up view of the *1101's* painted on identification..."*Me - P1101 V1." The* machine's fuselage appears previously damaged by *Messerschmitt AG* workers. For example, the fuselage immediately above the starboard wing root shows a great deal of damage. It is likely that this photo was taken of the *1101* after it was brought into its former workshop after spending time as a static display outdoors.

In the foreground is a *E-4* flying bomb with the *Projekt 1101* in the background. U.S. Military Reports stated that the *E-4* was a wooden mockup and not the real thing.

The *E-4* and the *Projekt 1101* from another angle in the workshop at Oberammergau.

The *1101* and *262* on static public display duties at Oberammergau about mid 1945. Both aircraft are suffering as they have fallen back on their tails.

The new appearing *Me 262* without its *004B* turbojet engines found inside next to the *Projekt 1101* by U.S. Army troops at Oberammergau. Like the *1101*, it too was moved outdoors to serve as a static public display.

This how the *Projekt 1101* suffered much of its fuselage damage. Pictured here is a American *GI* standing/sitting on the port side wing root while being photographed for relatives and friends back home in the United States.

This is the *1101's* physical condition after being relieved of static display duties. It is back inside its workshop along with the *262* at Oberammergau at the request of *Robert Woods*, chief designer at *Bell Aircraft*, Buffalo, New York. Notice the flat port side main wheel and nose wheel tire.

A close-up of the *1101* after being brought in doors. This photo is of its starboard side and it appears that much of its fuselage-skin damage immediately above the wing root has already happen and not when it was damage in a so-called "handling" accident back at Wright-Patterson AFB, Dayton, Ohio. This "bird" was never destined to fly before it ever came out of Oberammergau, Germany.

The *Projekt 1101* has been placed in a wooden cradle after being brought in doors. It does not appear that a great deal of thought and care has been given here...but that the aircraft was pretty much just lifted into its cradle.

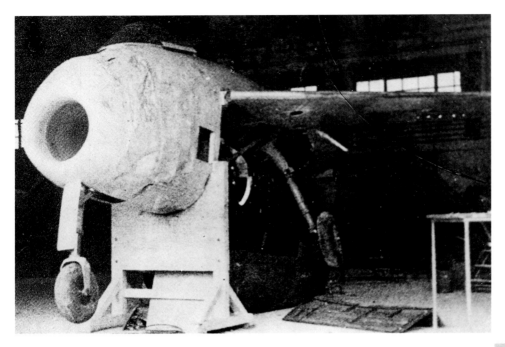

The nose port side view of the *1101.* Look at the fuselage skin damaged nose section back to the cockpit canopy and wing root. It appears to be pretty beaten-up from its outdoor static display duties where American *GI's* had the freedom to climb, sit, and walk all over the aircraft while on display. After all, no one from America wanted the aircraft...even *Robert Woods* came late to Oberammergau...after the *1101* was already performing display duties. It appears that he didn't have much left to work with once he obtained the machine.

A close up view of the *1101's* starboard side featuring its *HeS 011A* turbojet engine. It appears that the air intake on the *011A* is not connected to the *1101's* long air intake duct. It also appears that the *011A* shown is not in any sort of condition to run, either. At the top of the photo-center one of the three wing attachment points can be seen.

Woldemar Voigt's Projekt 1101 at *Bell Aircraft*. Metal workers are seeking to make the fuselage presentable, that is, to replace the most obvious damage. There is a good view of the underside of the port wing. It appears that the wing's under surface was not completed at Oberammergau as previously thought.

The *1101* at *Bell Aircraft* fitted out with an *Allison J35* turbojet engine in place of its *HeS 011A*. It appears that the *J35's* overall length and diameter fit the space available very nicely. Were the airframe in a condition to be safely flown, the *J35's* 4,900 pound static thrust would have propelled the light weight frame at a very high rate of speed.

Two *Bell Aircraft* engineers pondering the *1101* with its new American-made power plant. The height of the aircraft's wing pretty much allows an average man to walk under the wing without stooping. The diameter of the main landing oleo strut and attachment forging gives an indication from the placement of the engineer's hand. Note, too, that the under surface of the port wing shown remains open about the same way it was found at Oberammergau in mid April, 1945.

An excellent view of the *1101's* port side and wing main spar and ribs show with its *Allison J35* axial-flow turbojet engine. It is not entirely clear if the main landing gear would have retracted fully, however, given the *J35's* greater length and diameter from the original *HeS 011A*.

The *1101* all repaired, cleaned up, and painted and still carrying about its *HeS 011A* turbojet engine. About 1948 at *Bell Aircraft*. *Bell* engineers applied cannon templates (black) to the forward fuselage outlining the approximate locations where they amazingly believed the nose could hold three *MG 151* cannons per side!

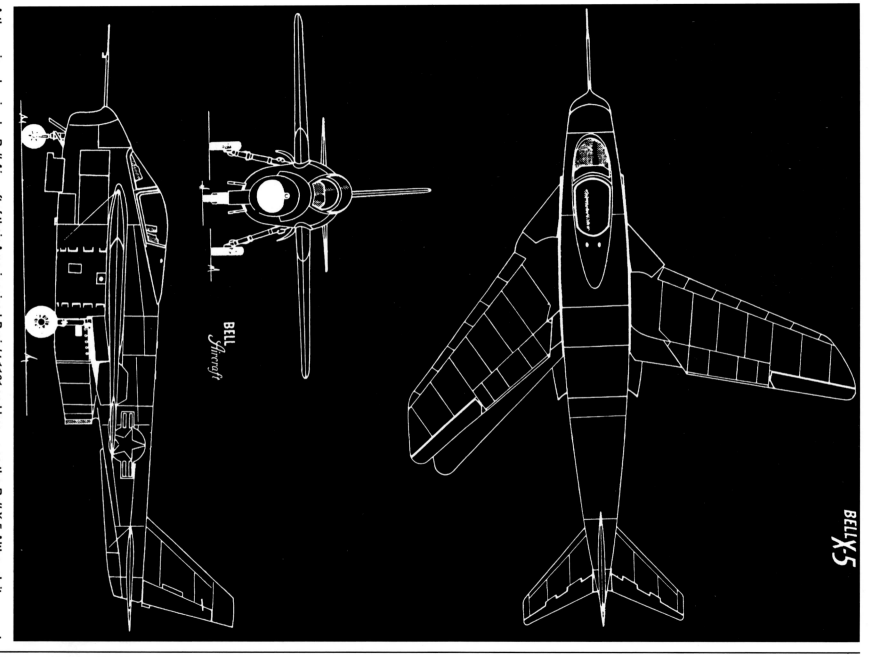

A three-view drawing by *Bell Aircraft* of their Americanized *Projekt 1101* and known as the *Bell X-5*. Although it may not be a clone, the *X-5's* superficial likeness to the *1101* is substantial.

BELL
Aircraft

BELL X-5

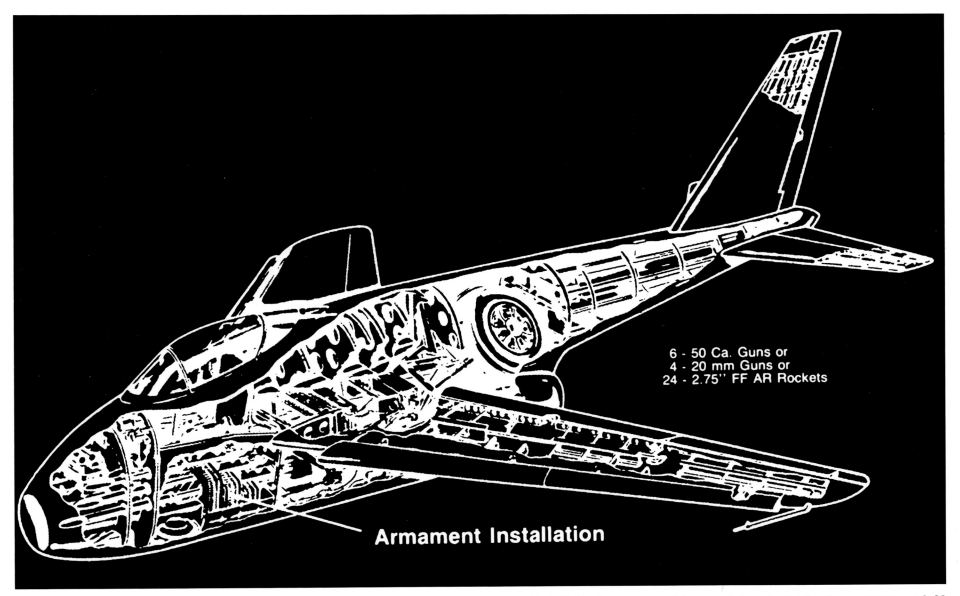

6 - 50 Ca. Guns or
4 - 20 mm Guns or
24 - 2.75" FF AR Rockets

Armament Installation

A pen and ink drawing of the *X-5's* armament installation. *Bell* engineers were proposing that their *X-5* had room enough to carry internally 6x50 caliber cannons or 4x20 mm cannon or 24x2.75 mm *FF AR* rockets.

A starboard side view of the *Bell X-5* during its flight testing at Edwards Air Force Base, California. Its flight testing proved out many of the alleged faults its predecessor would exhibit.

A front-on view of the *Bell X-5's* air intake opening/duct for its 4,900 pound static thrust *Allison J35* turbojet engine...looking almost exactly as the air intake on the *Projekt 1101*.

The *Bell X-5* on the ground at Edwards Air Force Base, California and because of the way it has been photographed gives a good view of the sweep back of its wing and tail plane.

Bell X-5. Revell scale model by Jay Miller/Aerofax.

The *Bell X-5* in flight with a full wing sweep back of 60 degrees.

The B*ell X-5* prototype #1 in flight high over Edwards showing off its starboard side with a wing sweep appearing to be about 20 degrees.

Lift-off of the *Bell X-5* prototype #1 (01838) at Edwards Air Force Base, California. This machine is now on display at the U.S. Air Force Museum, Wright-Patterson AFB, Dayton, Ohio.

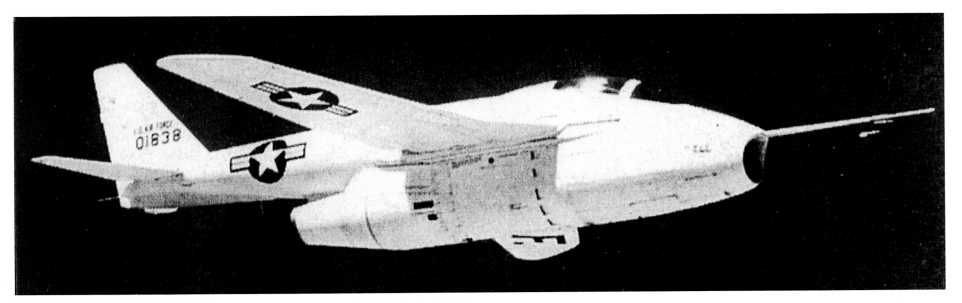

The *Bell X-5* prototype #1 in a low-speed flight and with wings swept at about 20 degrees.

The *Bell X-5* 1st prototype.

The *Bell X-5* shortly after lift-off showing how its main wheel/gear retracts up and rearward back toward the *J35 A-17's* jet pipe.

The *F-86* (left) and the *X-5* (right). Exhaustive flight tests confirmed that the *X-5* was not a match for the older *F-86* but the *X-5* prototypes made a significant contribution in knowledge and data aiding the design and development of other swing-wing fighters such as the *General Dynamics F-111*. Thank you *Herr Woldemar Voigt*. Thank you *Herr Willy Messerschmitt*.

A tail starboard side view of a camouflaged scale model *Me 1101* and giving a nice view of its complicated main wheel door covers. Jet exhaust from its single *HeS 011A* would have cleared the tail assembly by a wide margin of safety.

A port side nose view of a camouflaged *Me 1101* scale model and giving a good indication of the large surface area of its vertical stabilizer/rudder.

Red *17*, a *Me 1101* variable swept back fighter/interceptor is shown armed with four *X-4* rocket-propelled air-to-air anti-aircraft guided missiles. Digital image by *Mario Merino*.

A nose port side view of a *Me 1101* scale model camouflaged in full battle dress. Offensive armament would have included *Mk 108* cannon and up to 4x*X-4* anti-aircraft air-to-air guided missiles.

This *Me 1101* shown about dawn is being propelled by a single 2nd generation axial-flow turbojet engine...the *HeS 011* of 2,866 pounds [1,300 kilograms] thrust. Design specification called for a top speed of 605 mph [975 km/h]. Digital image by *Mario Merino*.

Red *17*, a *Me 1101*, as seen from its nose port side giving us a nice view of its two port wing mounted *X-4* anti-aircraft air-to-air guided missiles. Each *X-4* held a relatively large warhead of 441 pounds [200 kilograms] which was to be guided to Allied bomber packs at 550 mph [km/h] by means of a built-in homing device and then detonated within the bomber pack. Digital image by *Mario Merino*.